One Finch
Singing

One Finch Singing

poems

Emily Ransdell

Concrete Wolf
Louis Award Series

Concrete Wolf Louis Award Series

Poetry
ISBN 978-1-936657-91-9

Cover art by Anna Norris

Author photo by Steve Jones

Design: Tonya Namura using Corporate A Condensed

Concrete Wolf
PO Box 445
Tillamook, OR 97141

http://ConcreteWolf.com

ConcreteWolfPress@gmail.com

Table of Contents

III.

One Finch Singing

I.

Night Sky

It goes on forever,
beyond error and indignity,
accusation and bruise.
No matter how many sorrows
are torn from memory, night
shines its thousand lights down.
The archer, the bow,
the bull's bright eye. The dipper
on the back of the bear.

Once I was a girl who rode
the long tails of wishes
as I lay on the dock looking up.
Ice and iron is what
they were made of.
Beauty and dust. Night
like a kind of forgiveness,
letting the flaming pieces fall.

By Way of Introduction

My mother called me Petunia
                though hers never bloomed.

I was part empty flower pot, part lead paint.
                She was thumbed-through,

magazine pages of do-it-yourself projects
                left unfinished, a cotton housedress worn

unhemmed. She drank Jack Daniels from a juice glass,
                ashtray on her knees,

lawnchair under trees. Swelter of summer night.
                That's where she was when her water broke.

Four weeks early, drunken bees careening
                through the spindly peach trees left unpicked.

I shoulda named you Elberta, she said.
                All my life I've had to hear that.

Sometimes she called me Sweetie Pie. Said you can tell
                a lot about a person by the kind

of pie they like. *Take you,* she said
                one night, a little juiced.

You're the blackberry type.
                Your perfume alone is praise.

You're what I'd say if I prayed.

O, Vanity

The Rexall was nearly empty,
cashier distracted, the cosmetics aisle a gauntlet
of promises.
        Hot Lava.   Rulebreaker Red.
By morning, I was sorry.
Not for what I'd stolen, but for the kitchen's
cold silence, my lips reddened for my entrance,
then the burn of my father's scorn.
        Silent Vice.   Pink Sin.

What is so wrong with wanting to be beautiful?

As Kay was being vanished by chemo,
we talked about eyebrows,
lamenting the loss of hers. Without them,
her face wore the blank expression of a house
unoccupied, shades pulled down.

She said my own brows looked harsh, too dark
for my silvering hair. I thought then of my mother's,
overarched and unsteadily penciled, unnaturally
auburn to the end.

O, vanity. Threshold of pride.
Three days before dying, she woke
from a coma to ask for a touch-up.
Just the roots, she said, *a shampoo and set.*

My mother. Highballs and Noxzema.
Max Factor and Virginia Slims. Saliva
with the stench of toll-road coffee
spat on a Kleenex to wipe my chin.

The Buck

His father taught him to dress it
in the field, to whet the blade
and core the anus, clip the balls
then slit the hide. Run the knife
pelvis to breast.

I watched him straddle the split thing,
struggle it off the tailgate, dodging
the rack. He was sheened
with sweat, desire as plain
as the strain of each heft.
For a truck of his own, a job after school,
the impossibly soft hands of a girl.

Fifteen, he seemed a man,
shoulders lit by the street lamps
of our cul-de-sac, October's enormous moon
looking down on the houses where our parents slept.

He said when you cut the windpipe right,
the innards slide out with a single pull.
Heart and liver, lungs and stomach,
all of it linked like pearls on a string.

The buck hung over the tailgate,
neck bent back, chestnut eyes
gone dull. Some boys love death
more than anything. Some girls love
to look.

Bless the town

that loved me, the mothers who watched
behind home-sewn curtains, who knew
my whereabouts when my own

did not. Bless the playground and park,
the picnic pavilion that nightly drew us
like moths to light. Bless our initials

carved on the table, the pilfered beers
and sloe gin. Bless Ohio. Before meth
and Fentanyl, factories still cranking out

appliances and headlights. Bless
the class secretary and the running back,
that boy whose name I can't remember now.

Bless the back roads we drove for kicks,
two-lanes laid between corn fields
so straight you could cut the lights

and drive through the dark on a dare.
So straight you could roll a joint
while you steered with your knees.

Wild asparagus grew in the ditches.
Someone always had a pocket knife
and everyone knew the rule.

You needed to leave a few stalks
standing. It wouldn't grow back
if you took it all.

As If

We had nowhere to go but his father's
double-wide, its waterbed and faux wood
paneling, gridded asbestos overhead.

I watched him slip off his coat, then kick
off his shoes, step from the bell-bottom jeans
embroidered with peace signs.

I had never seen a boy without clothes—
his ghost of a torso, ivory hips. He looked
carved from marble, something priceless

I had been told not to touch.
He guided my hand to his penis,
its sweet velvet tip.

I lay adrift on that waterbed,
beer-buzzed and queasy, as if
I didn't know about consequences.

Already I was someone
who hungered, though I didn't know
for what. I didn't know about tenderness

or rapture, the simple epiphany
of touch. How much
could be mistaken for love.

Yes, I would go back again,

past the paved-over towpath,
the dry canal bed where barges
once rode low through the locks,
tobacco leaves with their scent
of earth and sweet briar bundled
and stacked on the dock.
Across the Great Miami, silted
with skunk weed and musk,
past the brickyard's last chimneys,
the chain-link rusting around
the husk of Carrolton Box.
At the top of the hill, I'd shift
into low like my father had shown me,
coasting past hawthorns shining silver
in the headlights, thickets still hiding
the dangers of darkened cars.
He'd be in the seat beside me, alive.

The Visit

When I knelt to face him, he said my name—
not the one he and my mother
had given me, but the long-forgotten one
he had called me as a child: *Lem*, he said,
so matter-of-fact. As if half a century
had not passed—no minnows, no fishhooks,
no father chasing a porcupine with a towel.
His precision like a surgeon's then,
the delicate untangling
of threads, removing without breaking
the prize of six spiny quills.

I wheeled him back to his room as he asked,
although that was not what he meant.
Back through the hallway gauntlet of slack
and spittle, past docile women in open-backed
shifts left unsnapped for easy hygiene,
past the unmistakable smell of shit.
To his room with its narrow bed like a child's.

It took two kind aides with arms like oars
to lift him. They were men
on a schedule, so many to tuck in
by dark. Still they paused
for the briefest moment as he settled.
I watched from the doorway
as the sheet floated white
and quiet. They waited
like parents do
before turning out the light.

My mother sits wordless

beside me as we drive to the hospice
where my father waits to die. I see her now
as she had looked at thirty, lost
in thought at her desk, hammering away
on her Underwood, a white snake of smoke
rising from the ashtray just out of view.

I'm a little lost myself, so exhausted I'm slow
to notice as the car ahead begins to drift.
A Buick, unmoored. Someone old, I suppose.
I say, *Get out my phone*, but my mother fumbles
and can't turn it on.

What if that driver is dying?
I think of the day my own heart
went haywire, my senses sucked one by one
though a great portal at the back of my head.
Gone was birdsong, then the dank scent
of river. Around me, cedars and firs,
riverbank dimmed as if dusted with ash.

Even then I thought my job was to save
everyone—my father, who panics at sundown,
my mother who needs dinner by six
and her cocktail. My mother who can never
fix either alone. In front of us now,
the Buick lurches and slows.

I want to weep.
I want to watch from a distance as though
my life is someone else's, to wave

as my parents sail away on a cruise to Alaska.
I want to see again those handsome
paramedics running to save me.

Mostly, though, I just want to weep.
Not for my father, so close to forever
I can only wish him godspeed. Not for my
mother either, or the clock she drew
for the doctor, its unintelligible hour.

For what, then? The slow-motion collision
that's surely coming, the Buick barely moving,
the driver's chin now dipped to his chest?
It's almost noon. Soon my father's after-lunch
morphine will kick in and we need to get to him
before then. I don't look at that driver again.
My mother doesn't look at anything.
Not the flaming roadside maples or the wreckage
of clouds in the midday sky. We pass the Buick
and beeline to my father's side.

Vigil

Outside his window,
the August night steams.
Parking lot mostly vacant,
              a black lake of heat.
Further out, a faint glow—

someone from the night crew out for a smoke.

At the lot's far edge
loom silhouettes of linden trees where earlier
I had found solace.
              Then they were alive

with bees, deep vibrations quaking the blooms.

Now such quiet.
Death is taking its time.
              Don't tell me

bees must return to the hive.
I know what it is
              to be weary.

Let them rest there till morning,
              when daylight will rouse them

to finish what they've begun.

Leaving Ohio

What did I lose along the gravel
pits and pastures, the scent of pig iron
and sulfur in the wind? Asphalt skies
above foundries and bean fields,
August heat.

Goodbye to guilt. To my father's
bones encased in lead. Goodbye
to his bucket list, found in the drawer
with his razorblades and aftershave.

Memory hangs heavy over everything
here—half-finished bird houses,
planters unassembled in the shed.
The pioneer gristmill and the towpath,
the ancient Indian mound rising
from the flatlands out by the golf course,
no artifacts ever unearthed.

Goodbye to its hundred shallow steps
I ran each morning as my father
drifted in his hospice bed.
Goodbye to the gods I prayed to
when I got to the top.

Salt

Born of sea or earth,
boiled from brine or dug.
Once carried by camels across the Sahara,
once carted by slaves on the road to Rome.
Salt for barter.
For cedar from Phoenicians, for glass
and the coveted purple dye.
Once traded for gold, weight for weight,
paid to sailors and soldiers at war.
When my father walked
death's long furrows,
I breathed the salt of his effort,
touched the dust it left
on his pillow and skin.
Jesus said, *you are the salt
of the earth*, and praise was born.
What could be more essential?
Salt livens the soup and brightens
the bread. Passed hand to hand
around the table, harvest of sorrow
and labor on our lips.

Finding a Long-forgotten Letter

Daisy's Cafe was the wrong place
to tell them, wrong time. November,
dark by five, a threat of black ice. We needed
to hit the road. For two days I had stalled,
afraid to ruin the mood of our fine family
weekend. They liked him, had taken
such care to fix a bed for him in the den.
We shouldn't have told them
at a cafe, should have braced ourselves
better for the face-off across that Formica table,

coffee bitter from sitting so long.
No, we had said.
We have no plans to marry.
No one spoke
of our family's faith or anyone's moral
compass. No need. Still, I could not
have imagined the letter that would arrive
a few days later–not the admonishment
I expected, but lament, a sorrow deep
as death, written in grief's own
hand, and in stone.
That winter, blizzard after blizzard,
they never called.
I made meat pies and stews
with stringy chuck roasts I got on sale.

One morning I found a frozen cardinal
on a crust of snowbank, bright
crimson against blinding white.
I missed my father.
My mother. News from home. I missed
that bird in the snow even though I had never

seen it alive. Impossible that such beauty
could be dead. I was young.
Righteously in love
with love. I can't imagine why
I saved that letter. And how
in God's name it surfaced after all these years.
After we married,
I had thought myself forgiven.
Yet still I stayed loyal to my shame.

Glenwood

In three days, she'd be dead,
but on Monday my mother rallied,
woke from her coma and asked
for a shampoo and set.
After that, she sat by her bed
for dinner, hair freshly curled,
the cannulas removed, for a while,
from her nose.
On her tray, string beans
and meatloaf, a slice of pie.

For the longest time I watched her,
followed her fork's slow journey,
its tremulous lift, then the long pause,
food falling back to the plate.
When I could bear it no longer,
I reached over, loaded her fork
with pie. Scent of cinnamon
and soft autumn apples,
crust like heaven,
a dab of whipped cream.
I raised it to her mouth, which opened
translucent as a bird's.

She chewed and swallowed,
then opened again.
The years remained unforgiven.
A mother. A daughter.
Alone among caregivers
and paid attendants,
the hovering nurse.

At night I hear her

rattling cake pans and candlesticks,
        rummaging through her collection

of souvenir tea towels
        from national parks.

She wants to know what I did
        with the pictures of her young

and pretty, why I sold the trunk
        with her wedding dress still inside.

Where, she asks,
        is her ceramic umbrella stand,

the champagne glasses etched with bells.
        Where are the Christmas dishes,

the cast iron skillet, the crystal punchbowl
        with matching cups?

Where is the cornucopia that sat
        on the dining room table, spilling

its plastic fruit? She wants it all
        back and buried

beside her, white carnations and a fifth
        of Jack Daniels left at the grave.

Flying My Mother Back to Ohio

After the plane rose from the tarmac, I sat
in coach with a drink. Rivers grew small
beneath us, bright sparks of broken-topped
cottonwood, Oregon's immense oceans
of green. I didn't think of her coffin
resting in cargo or wonder if the photo
and lipstick I sent had been of any use.

We crossed the ragged spine of the Rockies,
engines in the headwinds droning like the heart
of a weary white bird. The Great Plains
splayed open. I ordered another gin.

I had hauled out the mattress filled
with piss stench, scooped up the bathroom counter's
deadly confetti of meds. Met with the undertaker
and the auctioneer. The realtor, the church,
and the bank. Called 1-800-JUNK.

At the viewing, I saw that someone
had painted her nails Pearl Beige—
or maybe Rosy Taupe—one of those
brownish shades she loathed.
Shouldn't they have asked?
I know so little about protocol.

I had said thank you for the pie
from Costco, the casserole and coffee cake,
all of which I ate until I grew numb.
When I watched her body go
down I did not think
to toss in a carnation or a handful of dirt.

My mother was a bitter woman.
Did everyone guess I was glad she was gone?
I knew so little about forgiveness,
whether it could cross the fragile border
between the living and the dead.

Bowling in Heaven

Like newlyweds,
my parents slip out of their clothes.
He puts aside the sweater I chose
for him, she undoes her pearls.

They rise up from their old ailments,
their fears of falling, broken hips
and other bad news.

Now they dance
barefoot in their living room,
go bowling on a whim.

They garden all day without pain,
calling out like songbirds,
come see the hollyhocks,
they have grown so tall!

Nights, they lie down
like dolls and their sleepless eyes
glide closed. They seem so eager
for morning, I pray they find
each other again.

II.

One Finch Singing

Some days I want to fill my pockets
with everything I'm afraid of losing.
How much milkweed to save
the monarch? How many foil blankets
to keep an ancient redwood alive?

I worry about finches. Smaller
than a fist, wingspan no bigger
than an open hand. I keep thinking
of what it took for them to get here, flying
all those miles up to Oregon. I keep thinking
of heat. Cities hitting triple digits. London
for god sake. Italy on fire.

There's smoke again in Ashland,
like the time Kay and I went for a getaway.
All we had were bandanas, useless
against that stench and ash. We walked
the streets like grandmotherly bandits,
drank gin with the Airbnb windows shut.

By then I knew she was terminal.
Still, it felt impossible she could die.

I worry about beetle kill and rivers
missing their fish, the dry tinder of California
as creeks in Kentucky rage.
I read that finches can live on thistles, as if
to say, *There's hope.* The ancients thought

finches carried souls to the afterlife, and the sound
of one finch singing meant an end to grief.

Last week a brush fire ignited within sight
of my porch–*just like that*–flames leapt
from slash and grass to standing firs.
Two thousand acres burned.
Where did the birds go then?

I miss my friend.
I want to know those finches are somewhere.
Safe and singing. From meadow rush
and ditch shrubs, calling to their kind.

Absence

is like holding someone's heart
in your hand–dense
and heavier than you'd expect.
Like trying to speak
an unfamiliar language,
mouth opening
and closing like a door
on a silent hinge.
You're alone

with your one little ladder,
the whole orchard making
its fragrant demands.
Or you're plowing a field
full of rubble, dull blade
tolling halfway to heaven,
stone after ringing stone.

O, the stillness at the end
of each furrow, the mind
like a bell with no clapper,
no need now to pull the rope.

Nehalem Bay, No Moon

I'm back at the boat ramp where this time
last year we watched that huge moon
crest the coast range, how it lit
the ghost-colored stumps above Wheeler
like stars. The whole town glowed like a postcard
and we agreed there couldn't be anyone alive
who wouldn't want to be right here.

Tonight, no moon at all.

Where last year shoreline lapped,
driftwood sours in muck.
Tomorrow, chemo. The port
already in your chest.
In a vein called the vena cava, which sounds
so lovely, like a fine wine or the inside of a star.

Vena cava—pathway to the heart's
upper chamber, its atrium,
the moonless courtyard of this house
you have entered and now must walk through,
room by room.

When she said it was terminal

I became that animal,
the one I once saw dazed

by the roadside, jack pines behind her
charred to black bone.

How many summers will grief require?

Some days feel like forever, a slow
pirouette of sorrow spinning

like a maple's winged seeds, conjoined
as they are to travel further,

as if to better the odds.

Ode

To the Sunset Highway climbing
the Coast Range, crossroad towns
of Mist and Jewel, grass bowing down
in ballfields, petered-out farms.
Ode to this diner, halfway home,
your voice on the phone.
Ode to the waitress, stern
as a mother when I take your call.
To the words *lesion* and *seizure*,
and the upbeat note you ended on:
treatable with radiation,
though all I heard was: *it's in my brain*.
Ode to time lumbering by like trucks
on the highway, the concrete sky
of another Oregon winter bearing down.
To the cat the coyote got,
the dog you had to put down.
Ode to here. Ode to now.
Vials to be emptied, dressings
to be checked and changed.
Ode to everything.
Another round of surgeries,
another round of indignities.
Ode to the unknowable future,
a house with its doors open wide.
A woman at the threshold
readying to step outside.

Elegy, Interrupted

April 2020

I obey the arrows
of the grocery's one-way aisles,
keeping six feet away from anything
that breathes. I try to look
bright-eyed on Zoom.
How are you doing? asks everyone.

To grieve for anyone is to grieve
for everyone—I'm ashamed
to tell you I have not cried.
To kill time in lockdown, I scroll
through photos—Christmas
after Christmas of us looking older,
summer after summer,
more sensible hats.
These are the days we mourn
without ceremony.
Not even a thrift shop open
to take your clothes.

Year Two

The Times is asking for reflections
on pandemic grief:
     What have you learned about loss?
     Be as specific as possible.
Who hasn't missed the clatter
of restaurants or the laughter of strangers,
no one nervous or afraid? I miss
chatting about nothing, drinking wine
with my friend. I miss the way I once noticed light
at the cusp of seasons, which is to say I miss

predictable change. Now from the kitchen window
I watch the same wan shadows lengthen
then retreat.   Fencepost.   Flowerpot.   Firs.
Even the sun seems unsure
if it's spring.

This time last year sorrow was merely
starting, yet the garden grew on.
The limbs of our prize viburnum
hang heavy now, burdened
with buds, blossoms gathering strength
to shatter us again.

Three Dozen Tulips

Just weeks before lockdown
we had met at a restaurant, then walked
to Sephora, where they taught her to draw
eyebrows that looked real.
They look good, I said.
They go with your wig.

The morning before her death,
I left three dozen tulips on her porch step.
Sidewalk flanked by lawn
that needed cutting, sodden Lenten rose
and spent azalea debris.

I saw my reflection in the storm
door, blank as an apparition in my homemade
mask, my Safeway tulips a ghost bride's
bouquet. Inside the house, she was dying,
yet I had brought the most garish flowers
I could find. Canary yellow, cherry red.
Like confetti or birthday balloons.

What was I thinking?
In her garden, Lilies of the Valley hung
their pale heads in the sun.
All over the city,
things were shutting down.

How to forgive

the crickets and bickering crows beneath
the bedroom window, the low-slung
stalk of the cat with a hummingbird caught
in her mouth. The sightless moles
that bulldozed the grass as they built
their dark basilicas beneath us,
altars of roots and grubs.
Cabbage worm and slug, cosmos
grown leggy, day lilies gone.
We had shared a bottle of Albariño
on her back step and she asked
if I'd help her shop for a wig.
Now ice bends the fenceline
junipers, cold rain lashes the firs.
Our friends will gather
for Christmas dinner as always–
pecan pies and a hothouse poinsettia.
I'll bring the same
champagne as last year
and a centerpiece
for the table, boughs fresh-clipped
from the backyard holly tree,
bright with red berries,
bright with sharp shining leaves.

After a Freak Tornado

Body of sorrow,
body of sweat and prayer, body
of panic and ravenous want.
My lungs open
and close with their hungers,
my breath rises
and falls like heavy wings.
More like a body inside
my body, breathing me.
Out on the street, the shock
of debris, cottages stripped
of shingles, massive old hemlocks
twisted and split.
I cannot look out the window
for comfort now,
with all the branches gone.
Show me how to grieve.

In August

                    All day the cedars
have been dropping
needles, yellow
and brittle on the deck.
                     So dry you can hear
them hit,
even inside the house.
                    They sound like rain,
not the dark, familiar drizzle
we crave,
but the drenching kind we need.
                    Hemlocks parched,
spruce and shore pine
tinged the color of flame.
                    Dorianne says
if trees could speak
they wouldn't.
What's there to say anyway—
                    days so hot,
talk is just more tinder for fear.
                    The Bootleg fire
was the size of L.A.,
                    then Rhode Island, twice
the size of Manhattan
where the Statue of Liberty stands
                    gowned in smoke.
All the way north to Nova Scotia,
people are posting
photos of blood-red sunsets,
                    West Coast wildfire haze.
But here in Manzanita
our air is clear.

The scent of disaster drifted
                    elsewhere this time.
Behind the house, bracken ferns curl
beneath the laurel as if seeking
                    shelter. Even the squirrels
have gone quiet, even the crows
                    and the jays.

Driving Through Idaho

This road is a ribbon of unspooled
      loneliness, rimmed with the tinder
      of endless grassland,
farmhouse flower patches parched
      to a passing blur.
All along the highway,
      hand-lettered signs say *Bless*
      Our Firefighters.
Waving children offer water and snacks.

Out by the Walmart, a makeshift mess tent,
      sandwiches by the hundreds where everyone
      hungry is fed.
When I get home I should try to be kinder.
      I should hold open doors for mothers
      with strollers, help the elderly reach tea bags
from the grocery's highest shelves.

I should remember Idaho,
      these dung-colored skies, the cooler
      of sodas left at this crossroad, ash-covered
bills in the Folgers can taped to its side.

Early May on the Pitt River

Already above the sugar pines
    clouds look threadbare, casting
        their pallor on the canyon oak
           and the doe who grazes

on nothing. This morning I saw her nosing
    an abandoned campfire ring.
        Broken glass.
           Charred cans.

The saddle of her shoulders exposed,
    she is an echo of the river,
        water already too warm
           for trout,

bone-colored bedrock bared.
    She drinks and I watch her haunches
        quiver. Her eye on the river, the river
           a sky with its eye on her.

When Carol reads her poem

with its phrase, *earth's late afternoon,*
I think of birds on their way
elsewhere—geese in efficient
formation, pelicans the color of mud.
Gullets and the great engines
of their wings unevolved over millions of years.

For the first time in anyone's memory,
they've abandoned eons-old flight plans
for the refuge of farm ponds, spring-fed
and clear despite drought. Carol reads
earth's late afternoon and I think
of scorched caneberries. The hell-summer
scent of smoke.

Winter is coming.
Quince and crabapple trucked to market,
the last of the pole beans drying on bamboo.
What migrates will go.
Carol herself is leaving, her husband
is dying, and they've sold their farm.

She reads and I think of spiders
nesting in the toolshed, gapped floorboards
welcoming the wintering mice.
Someone else's turn
to patch the barn roof now, to move
the pump from the creek bed
to higher ground.

What I Know About Fire

It can jump a river.
Embers can travel twenty miles
to land on a porch.
Night after night I watched
the moon like a copper coin rise
above the skyline of cedars
and death-dry firs,
the smell of dirt and burn.
Now, finally,

rain. Roses battered,
chrysanthemums limp.
Petals swirling in the storm drain
like a frenzy of tiny boats
with no one at the oars.

I watch them spin
from the window I've opened
now that it's safe
to say I was afraid.

Summer of smoke
and uncertainty, all the wrong
colors in the sky.
When the yard dries out,
there'll be basil and mint
to bring in. The season's
finally quenched tomatoes
splitting their skins.

First Wind, Then Fire
Western Oregon, September 2020

Then smoke for days.
The air like a heavy animal, gray
and low to the ground.
Hot wind has stripped the viburnum
and flattened the hydrangea's last
blossoms, their massive violet heads.
A smell like charred rubbish
seeps through the house.

If I could sleep I might dream
of birds rising high in the thermals,
wings spread wide.
What has become of the raven
and the backyard jay, the red-tailed hawk
who frequents our fencepost, always
hunting what's helpless.

Even the moon has gone missing.
It's been a week since I've seen
the valley's far side.
Thirty-three fires are burning.
I don't know where
smoke ends and I begin.

Three hairless squirrels lie dead
in the driveway, newborns
blown from their drey. I cannot bear
to shovel them into the trashcan,
their eyes sealed shut behind
blue-veined lids.

They are the color of smoke,
the color of this sky wrapped
around Oregon, its grieving vineyards
blackened with ash,
alders that will never turn.
The crickets who've mistaken
this mid-day gray for dusk
and have begun to sing.

III.

Beloved

Bring me your fears. Bring them
like a handful of sad white lilies.
And your sorrow, bring that too,
in the walnut box your father made as a boy.
Corners tightly dovetailed, brass-hinged
heartwood varnished to a sheen, treasures
you left there decades ago still rattling inside.
Dust-colored sparrow wing, a cuff link,
the home address of that boy at summer camp
you couldn't save. Bring me the memory
of your high school sweetheart, the field
behind your house, the long minutes
you breathed for your mother until
the ambulance came. Bring me
your misgivings. Your heartache.
I'll haul it all to the river in a cart strung
with white carnations and won't ask
that you come along. You never did
like to talk about what's gone.
That's okay. I'll come back
with that cart scrubbed clean.

On the day of his biopsy

I'm up early, coffee on the sill
by the chair. Sky clamped shut,
the horizon's gray hinges locked.
        Across the street, another house
is humming, doorway a swarm
of crayon-colored raincoats and boots.
A woman bends at the kitchen window,
reaches, bends again.
        She's doing dishes, though
from here she could be dancing,
swaying between basin and drainer,
hands like white wings flashing
as she lifts each plate and bowl.
        A man appears.
He leans around her body, perhaps
slipping a coffee cup into the sink.
        I don't want that life
back, yet I cannot stop
watching. Any moment now,
the hive of that house will empty,
its windows going dark
as that family spills
into the world.

Unspoken

A word is elegy to what it signifies.
— Robert Hass

Forest, field, bramble, bloom.
Walks along the river, nights
gazing up from the backyard grass.

The way you always seem so cheerful
when you work in the kitchen,
the marvel of risen loaves in your hands.

Now we sit in a room
where your name will be called,

tables piled with ragged copies
of *Auto Week, Garden and Gun.*
Field and Stream.

Wife in the Surgical Waiting Room

Home's the place we head for in our sleep.
– Louise Erdrich

I was certain of nothing
until the night I heard you speak as I slept.
It was no dream. You said my name.
Rolled it like a stone
down the long hallway of years.
Later you swore you hadn't said a word.

Next morning an unfamiliar bird
sang in the junipers, so high
in the branches, the white underside
of its belly was all we could see.
That was a talisman too.

Do you ever wonder where we'd be
if we hadn't met when we did?
Each of us waking up elsewhere,
impossible even to imagine
how a day like this one would begin.

Portent

I won't forget how you looked
as they wheeled you
        to surgery, I won't forget
        waiting in the hall.

The heart is invisible inside
the body, a pulse undetectable
        across a room.

By summer this will all be over
and we'll sink back in our beach chairs
        to watch the Perseids
        plummet again.

Portents of fortune, diamonds
 above wheat fields and dust.
        Luck for shepherds and seekers,
        for weary farmhands toiling

even in sleep. Luck for monks
with their begging bowls,
        for mothers in the basements
        of bombed-out buildings.
Luck for lovers like us.

Everywhere a River

I do remember darkness, how it snaked
through the alders, their ashen flanks
in our high-beams the color of stone.
That hollow slap as floodwater hit
the sides of the car. Was the radio on?
Had I been asleep?
Sometimes you have to tell a story
your entire life to get it right.

Twenty-two and terrified, I had married you
but barely knew you. And for forty years
I've told this story wrong. In my memory
you drove right through it, the river
already rising on the road behind us,
no turning around. But since your illness
I recall it differently. Now that I know
it's possible to lose you, I remember.
That night, you threw that car in reverse
and gunned it, found us
another way home.

This River We're Crossing

Like the day we brought the baby home,
everything seems too bright,
too large and loud, this bridge a million
miles long. Then, I wanted to go back
to Good Sam, where a star magnolia
bloomed beneath my window
and the nurses wore quiet shoes.

Today, the same thundering semis, same
gorge wind shuddering the car. I'm afraid
to take my eye off the highway,
to ask if you heard what the surgeon said.

Below us, sun sparks the whitecaps
and sailboats, the picture-book lean
of their sails. Impossible from here

to see which way the current is moving,
impossible to tell how fast.

Oregon Nights

Before I climb the stairs
to sleep, I like to step
outside, lean against
the porch rail and breathe.
I like to listen as dark
rivulets of rain ping through
the gutters and long downspouts,
a thrum like heartbeat, rain
like lifeblood here.
Thirty years. I know better
than to search for stars.
Yet my gaze still lifts.
Above the black asphalt
of Ivy Street shimmering
like a river under
the streetlamp, above
the house across the street,
its one lit window
a little box of light.
Above glistening shingles
and the shadowed
firs. Forever the firs.
Their limbs feathered
with darkness, snagged
with remnants of prayers.

Arc

I watch the book rise and fall
on his chest as he sleeps.
I think of shorebirds lazing
on a river, a small boat
riding the current, drifting
past slack water reeds and sunlit
riffles to a shaded pool where
cutthroat wait out the heat.

That's where he's gone
with his long breaths, measured
like even pulls on the oars.
Nothing to keep him here
now that I've switched off the lamp,
set the book on the table by the bed.

With his eyes closed, he looks
ten or twelve, dreaming
the dreams of a boy.
Ahead of him still,
all those hours on the water,
learning the currents
and the light, perfecting the arc
of the line.

Threshold

I wish you'd seen the sky
                    last night–dipper filled
                              to the brim, the Milky Way
a curtain of pearls. So clear
                    above Wheeler we stood
                              on the porch in shirtsleeves,
pie uneaten, coffee cold.
                    Somewhere above us,
                              Orion pulled dawn's pale blanket
over his shoulders,
                    the seven weary sisters
                              turning homeward then,
toward daylight, and beds
                    of their own.

Spectacular

A chickadee flits outside our kitchen
window, tiny magician dressed

in sunlight and a sleek
black cap. It appeared from nowhere

and is hard at work. Does it know
the season is short?

Stand with me here, love,
where we can watch it build

its fragile house of sticks.

Aubade with Surgical Scar

At first, I was reluctant to look.
A scalpeled incision on your torso,
raw and red, sutured with precision,
but nonetheless *sewn*. Mended,
I thought. Needle and thread.
I remembered Hannah's stuffed bunny
caught in that long-ago car door,
the ceremony we made of my jagged repair.

This morning you open the blinds,
scar now silver, a distant river
crossing changed terrain.
Your silhouette comes alive
with light. Your thighs
and resolute calves, your backside
strong as a young man's.
And your ass. Two millstones
of muscle, one glorious
machine.

As a Hot Wind Pummels the Hydrangeas

I come out with clippers
to save a few pale blossoms,
carry them back to the house
like a bride. I lay the past on the table,
strip the rust-colored leaves
from their stems. I turn away
from worry and the photos of us young.
I bring in the garden hose and paper lanterns,
unhook the swing that knocks
against the side of the house. I stake
the lilac's tender branches and mulch the lilies
the way I've seen you do.
A storm is coming, yet I cannot bear
to trim back the fuchsia,
still in bloom.

O irrevocable / river /of things
after Neruda

I have rubbed your burl
with orange oil,

shined your tarnished handles
till they glowed.

I have starched the collar
of my father's burial shirt,

sealed my mother's wedding dress
in an acid-free box.

Each common plate in the cupboard,
each cup and cut-glass bowl.

What becomes of memory
without its possessions,

this wide river, such high banks
on either side?

The Typesetter's Daughter

The pig, the slug,
the Ludlow's liquid heat,
hot lead singeing the hair
off his arms.
The steel-toed shoes,
the apron
and ink, pocket stuffed
with paper clips and peppermints,
ball-point pens.

The lever, the lock,
the letters plucked
one by one from the case.
Her name he cast all caps
in Baskerville
for a paperweight.

The chair, the desk,
the pen. The voice
in her head.
Begin.

The canoe, the creek,
the calling loon at night,
the lake a waiting page
of light.

A girl at the margin
of that shoreline,
a color like silver
pouring itself into letters,
letters into words.

Once

I made a home of the dark months—
        like a thrush making safety
                from nothing but lichen and mud.
                    Winter was a nest

I built myself, lights on in every room.
        Sometimes I burrow even now.
                Drawn as I am to the scent
                    of wet duff, moss like mourning

shawls slung over firs. Sky a poultice
        of gray. This morning I saw a white-tail doe
                cross the road, nostrils steaming, head low,
                    her coat the color of stone.

One quick step into the understory
        and she was completely gone.
                It's foolish to think I know
                    what she's feeling—the trees

in silent kinship, their needled paths a labyrinth
        for forage and rest. Even the lonely are not
                alone in a forest. Human scent travels far.
                    Every animal knows where you are.

Heart Speaks

In waves of filaments flickering,
through pulsing valves like the mouths
of hungry fish. A fistful of hinges
that open and close the four doors
of heartbeat one hundred thousand times
a day. Heart like a stone dropped
in deep water. Heart like an anchor
drug across the body's unseen floor.
Heart that brought me to my knees
on an ordinary morning
walk, demanding its name
be called.
You have taken much for granted,
Heart says from its dark nest.
Yes, I answer,
yes   yes   yes.

The Tree

Rotorua Lake, North Island New Zealand

At the rim of a caldera filled once
with fire and the anger of gods,

I sat in a canvas camp chair
under an unfamiliar tree.

I watched black swans grooming
their broods in the shallows, swamp hens

squabbling for dominion, strutting
males with their violet breasts iridescent

in the heavy light. I sat a long while, certain
if I left I would miss something

important, as if the gods still had
a message to send.

In the capital city, the prime minister
had just resigned, the one we'd admired

for her aggressive legislation on guns.
At home, another random shooting,

another community stranded
in grief. Above me, the tree let go

a yellowed leaf. I watched it
drift down to the lake

now empty of color,
like the sky. Nothing

more happened. Not a breath
of air moved.

I mean to learn the name
of that tree, its trunk matted

by an impenetrable snarl
of vines. Dirt-colored and leafless,

it looked ancient, like an unkempt god
neither dead nor living, but dormant.

Waiting, like the other gods,
for its turn.

Kingdom

I am not religious
yet still God found me.
He named me Faithless,
whether or not it was true.
I stood before him, a stalk
in a wilderness of blossom,
unbending among verdant
and undulant vines.
I walked grief's dark fields
with my plow blade
sharpened, white gardenia
seeds in the palm
of my hand. God said,
What is your intention?
I licked the holy pollen from the air.

Acknowledgements

Thank you to the editors of the following journals in which these poems, sometimes in earlier versions and under different titles, first appeared.

Crab Creek Review: "Ode"

CALYX: "On the day of his biopsy"

The Cortland Review: "Bowling in Heaven"

Kestrel: "Yes, I would go back again," "Beloved"

New Letters: "Everywhere a River," "Leaving Ohio"

North Coast Squid: "Vigil"

Open: A Journal of Arts and Letters: "Night Sky," "Salt"

Pembroke: "Unspoken"

Poet Lore: "What I Know About Fire"

Poetry Northwest: "The Typesetter's Daughter"

Prime Number: "Bless the town," "Glenwood"

Rattle: "The Visit"

River Styx: "At night I hear her"

Ruminate: "Portent," "How to forgive"

San Pedro River Review: "Arc"

SHIFT: a Publication of MTSU Writes: "As a Hot Wind Pummels the Hydrangeas," "Spectacular"

Spillway: "Three Dozen Tulips"

SWWIM: "The Buck"

Sugar House Review: "By Way of Introduction," "Kingdom"

Tar River Poetry: "First Wind Then Fire"

Terrain: "Driving Through Idaho"

Turtle Island Quarterly: "In August"

Voicecatcher: "Flying My Mother Back to Ohio"

Whitefish Review: "Heart Speaks"

Willawaw Journal: "This River We're Crossing," "When Carol reads her poem"

Windfall, A Journal of Poetry of Place: "Nehalem Bay, No Moon," "Threshold"

"After a Freak Tornado" appeared in *The Practicing Poet: Writing Beyond the Basics*, Terrapin Books, West Caldwell, NJ. 2018.

"Beloved" and "By Way Of Introduction" were featured in *HipFish Monthly*, Astoria, Oregon.

"Beloved," "First Wind, Then Fire," "Unspoken," and "The Visit" received Pushcart nominations.

"Bless the Town" was runner up for the Prime Number Poetry Prize from *Press 53*.

"Bowling in Heaven" was reprinted by Ted Kooser in his *American Life in Poetry* series.

"Elegy, Interrupted" appeared in Essential Voices: *A COVID-19 Anthology*, West Virginia University Press, 2023.

"Everywhere a River" was featured in *American Life in Poetry* and anthologized in *SageGreenJournal.org*.

"Glenwood" was runner up for the Prime Number Poetry Prize from *Press 53*.

"This River We're Crossing" appeared in the anthology *Is it Hot in Here or Is it Just Me? Women Over 40 Write on Aging*, Social Justice Anthologies, Pittsburgh, PA, 2019.

Gratitude

Thank you to Concrete Wolf Press for selecting this manuscript as winner of the 2022 Louis Award and to its badass publisher, Lana Hechtman Ayers, who saw connections and possibilities and made this book whole.

For the richness poetry has brought to my life, I thank my teachers, mentors, and heart-friends.

Thank you, Andrea Hollander. My Pacific MFA advisors Ellen Bass, Kwame Dawes, Dorianne Laux, Joe Millar. Forever, Joe Millar.

Forever too, my poetry sister Jennifer Dorner.

My gratitude to The Suzannes. To Earl. Nancy, Jo, Jamacia, Carla, Carol, and AK. Michele. Dion.

To my students at the Hoffman Center for the Arts in Manzanita, Oregon. My gratitude to Vera, Andy, and Phyllis.

My monthly Zoom poetry pals Alida, Nancy, Allisa, and Eileen.

Thank you, Anna, for creating the finch I saw in my mind. And Ted, for showing me how to center the clay.

For love and support, thank you, Hannah.

For everything, thank you, Steve.

About the Author

Emily Ransdell lives in the Pacific Northwest. She holds MFAs from the University of Montana and Pacific University. Her work has appeared in *Poetry Northwest, Poet Lore, Tar River Poetry, Terrain, River Styx*, and elsewhere. Emily has been a finalist for the Rattle Poetry Prize and the New Millennium Writings Award, and runner-up for the Prime Number Poetry Prize from *Press 53* as well as the *New Letters* Literary Award. *One Finch Singing* is her first poetry collection.

CPSIA information can be obtained
at www.ICGtesting.com
Printed in the USA
BVHW030829110523
663996BV00002B/11

9 781936 657919